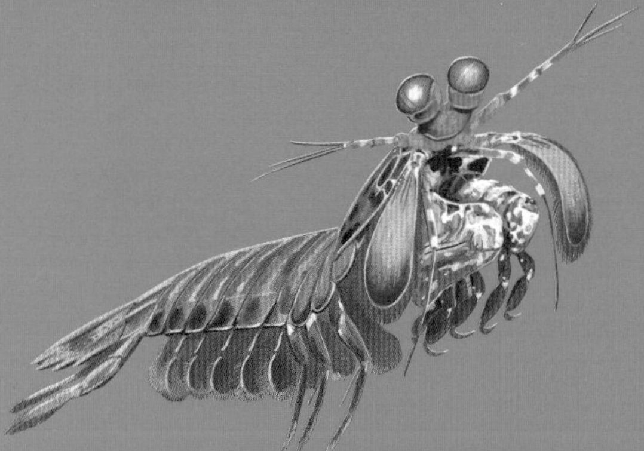

AWESOME OCEANS
ANIMALS OF THE TROPICAL SEAS

Michael Bright

Copper Beech Books
Brookfield, Connecticut

Produced by:
Aladdin Books Ltd
28 Percy Street
London W1T 2BZ

ISBN 0-7613-2815-7

First published in the United States in 2002 by:
Copper Beech Books,
an imprint of
The Millbrook Press
2 Old New Milford Road
Brookfield, Connecticut 06804

Editor:
Kathy Gemmell

Designers:
Flick, Book Design & Graphics
Ian Thompson

Illustrators:
Ross Watton—SGA,
John Rignall, Myke Taylor,
Ian Thompson, Norman
Weaver, David Wood
Cartoons: Jo Moore

Photo Credits:
Abbreviations: l-left, r-right, b-bottom, t-top, c-center, m-middle
4-5—NASA.
5mr—Jeffrey I. Rotman/CORBIS.
14-15, 18-19, 26bl, 28-29—Digital Stock. 25c—John Foxx Images.
Picture Researcher:
Brian Hunter Smart

Certain illustrations have appeared in earlier books created by Aladdin Books.

Printed in U.A.E.

Cataloging-in-Publication data is on file at the Library of Congress.

Contents

Introduction

The world's tropical seas stretch around the planet in a broad band on either side of the Equator. Here, in the shallow waters, there are mangroves, sea grass meadows, and coral reefs, which have a wider variety of marine life on them than any other part of the oceans. In fact, the coral reef itself is alive, and it offers shelter to thousands of sea creatures. Life in deeper tropical waters is limited to fast-moving predators traveling from one feeding site to another.

Spot and count!

Q: Why watch for these boxes?

A: For answers to the tropical sea questions you always wanted to ask.

Bits and pieces

These boxes take a closer look at the features of certain animals or places.

Awesome facts

Watch for these puffin diamonds to learn more weird and wonderful facts about the amazing creatures of the tropical seas.

Tropical seas

The deep tropical ocean is like a desert, but close to islands and underwater mountains, and in and around coral reefs, mangroves, and sea grass meadows, the water is teeming with life. There are so many different kinds of fish living there that scientists discover new species almost daily.

Islands and underwater mountains near the Equator attract sharks, dolphins, and many kinds of fish, such as the jacknife (above), to nearby shallow waters.

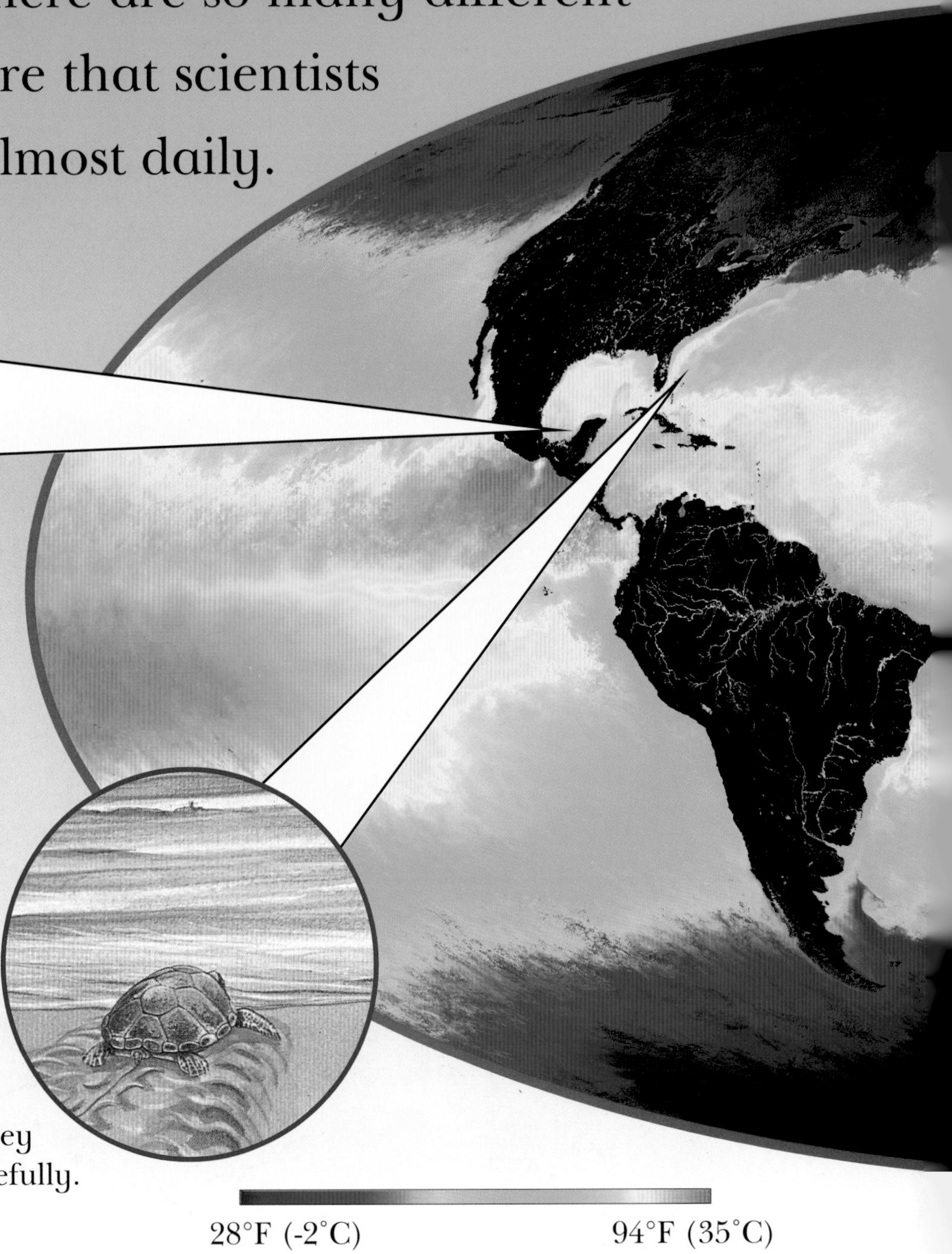

Sea turtles nest on Florida's Atlantic coast. They crawl up onto the beach, dig a hole, and deposit 100 eggs. When they return to the sea, they leave a series of telltale wavy lines in the sand, but they disguise the actual nest site carefully.

are bordered by cool currents.

Q: Why isn't there much wildlife in the middle of the tropical oceans?

A: There are few nutrients there, and plankton concentrates mainly at ocean current boundaries. Food tends to be "clumped" and many predators, such as swordfish, blue sharks, and dolphins, must travel miles to find it. When they find fish, they feed quickly, often herding them into dense clusters.

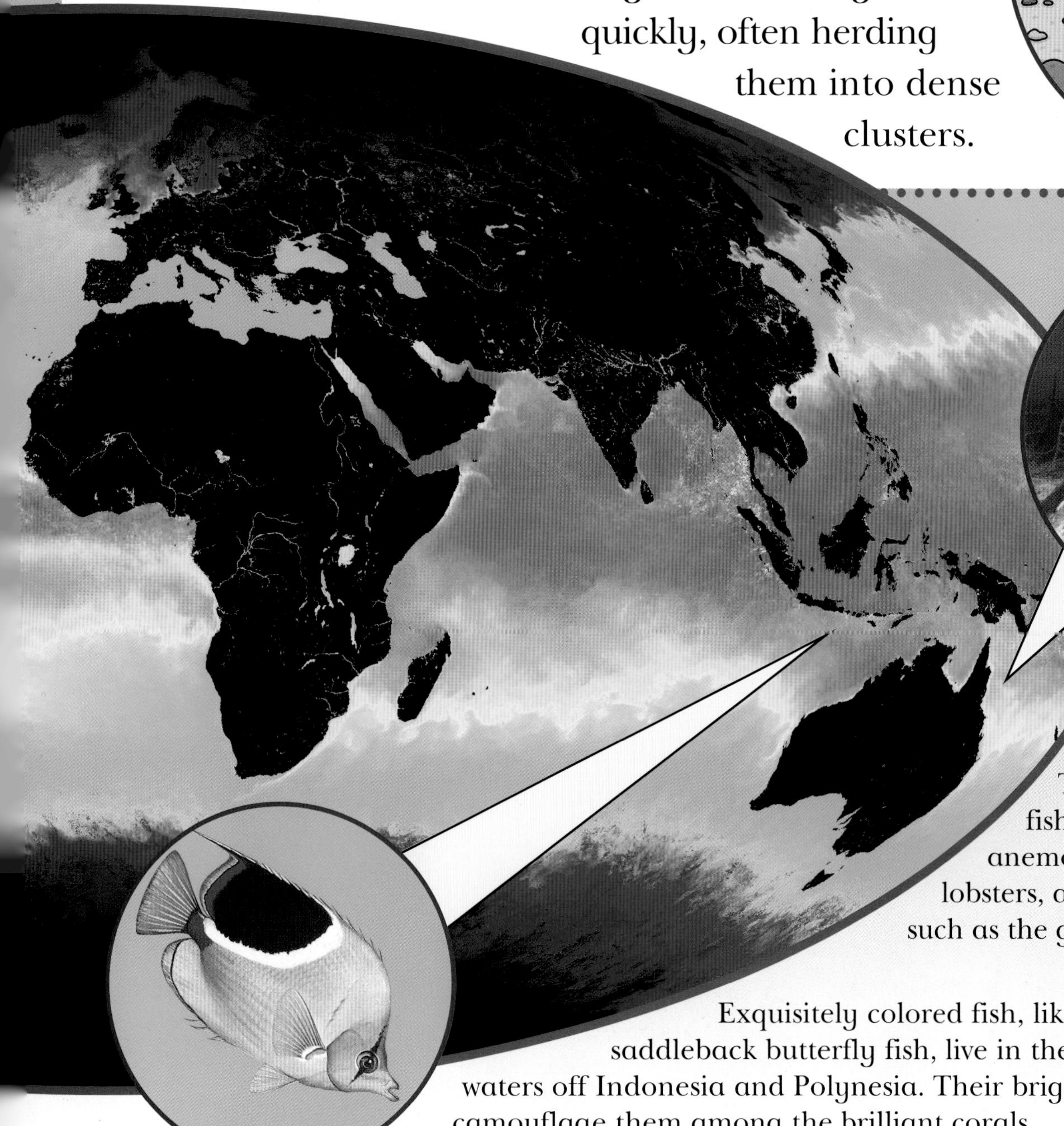

Coral reefs, such as Australia's Great Barrier Reef, are the gardens of the sea. They harbor many fish, jewel-like corals, anemones, shrimps, crabs, lobsters, and monster shellfish, such as the giant clam above.

Exquisitely colored fish, like this saddleback butterfly fish, live in the tropical waters off Indonesia and Polynesia. Their bright colors camouflage them among the brilliant corals.

Parts of a porcupine fish are highly poisonous, and people

Warm-water wildlife

All groups of animals are represented in the wildlife-rich fringes of the tropical seas—dolphins, whales, sharks, bony fish, crustaceans, sea stars, sea urchins, mollusks such as squid and octopus, marine reptiles such as sea snakes, seabirds, and a whole lot more. Most are colorful, with the color sometimes showing that the animal is dangerous.

Red crabs on Christmas Island in the Indian Ocean spend most of their time on land. But millions of females return to the water's edge on the same day each year to release their eggs. If they are caught by a wave, they drown. Their larvae develop in the sea. Then one day millions of baby crabs crawl out of the surf and up into the island's forests.

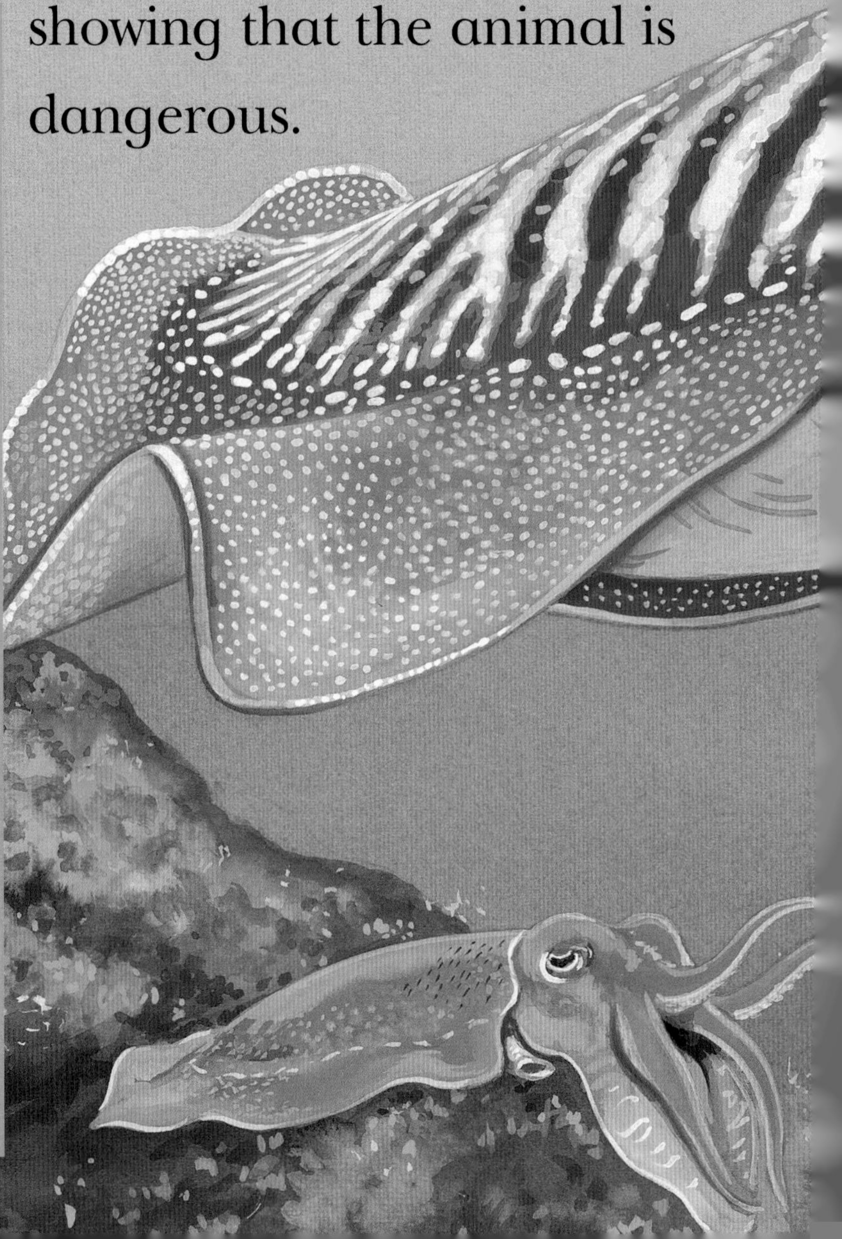

eating them have died.

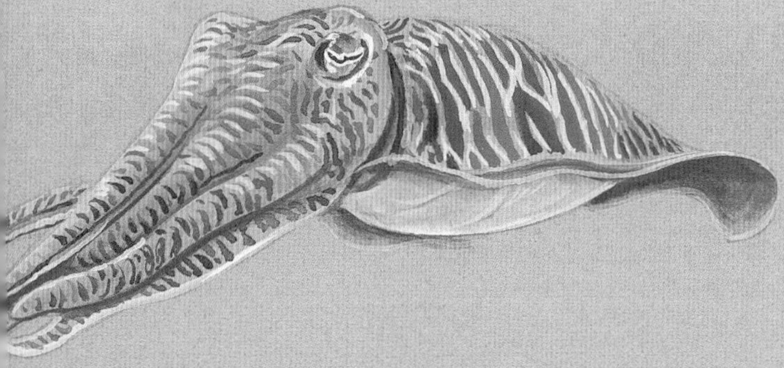

zoom in on...

Fish defense

The porcupine fish is a slow-swimming resident of tropical coral reefs. When attacked, it blows itself up with water like a balloon, so that predators cannot get a grip. Its sharp spines normally lie flat along its body, but stick out when the fish is threatened.

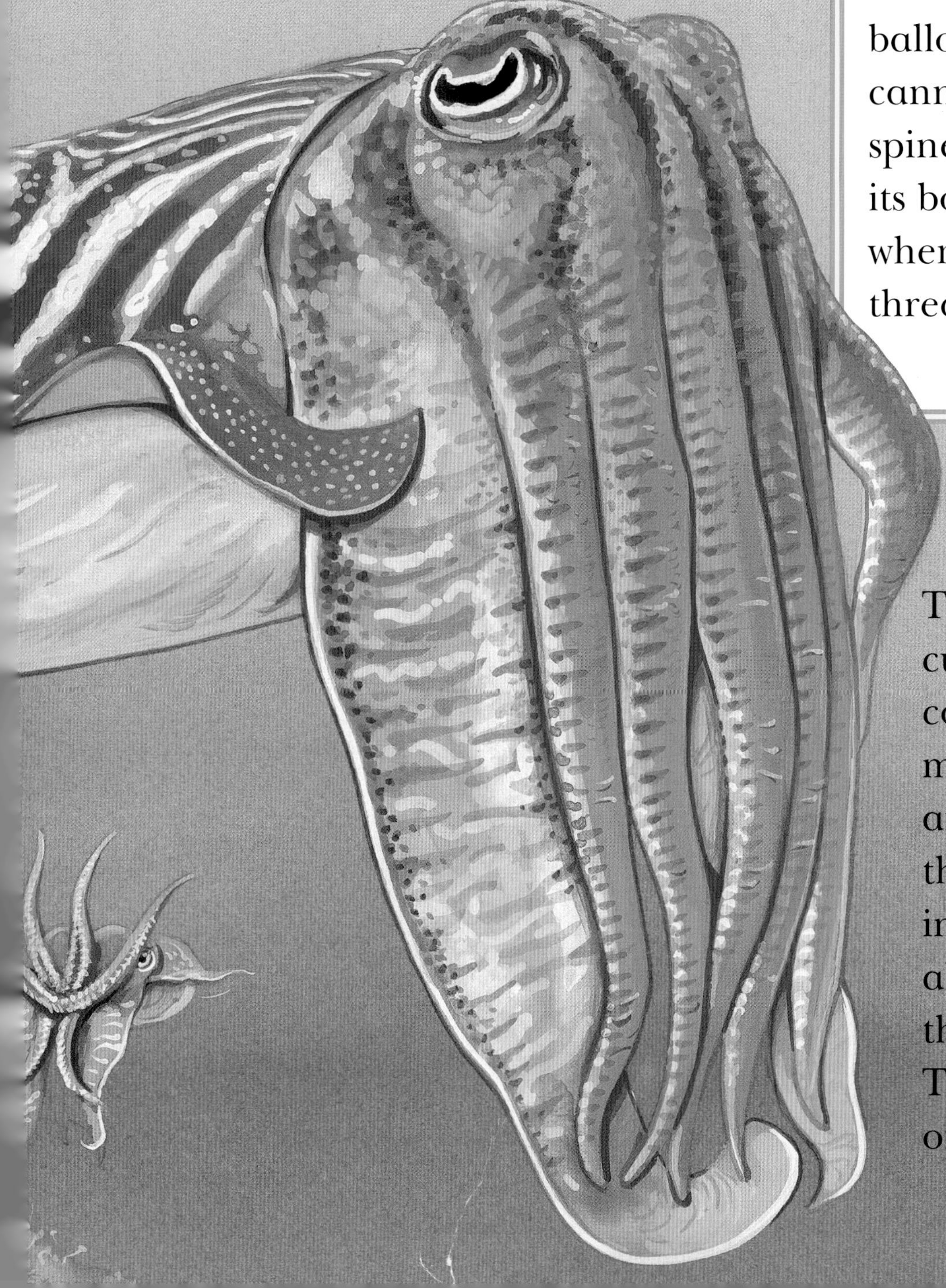

The skin of a pharaoh cuttlefish can change color in an instant, with moving patterns of stripes and v-shapes. It is thought that cuttlefish communicate in this way, but that they also use this to mesmerize their prey when hunting. The pharaoh cuttlefish lives only for about 240 days.

Corals with algae in their tissues grow 10 percent faster than

Coral reefs

Most tropical shallow-water corals have been built over hundreds of years by billions of tiny creatures called polyps. Different kinds of polyp have different shapes, resulting in many weird and wonderful corals. The largest collection of coral is Australia's Great Barrier Reef, which is an amazing 1,250 miles (2,000 km) long.

Q: How does a reef grow?

A: Slowly. Each polyp secretes a protective shell of calcium carbonate (lime). Live coral polyps live on the surface of shells left by earlier polyps. Together, the colony may add half an inch (1 cm) to the reef each year. Over thousands of years, this can produce reefs 900 ft (300 m) thick. Corals have green algae in their tissues, which supply them with extra food.

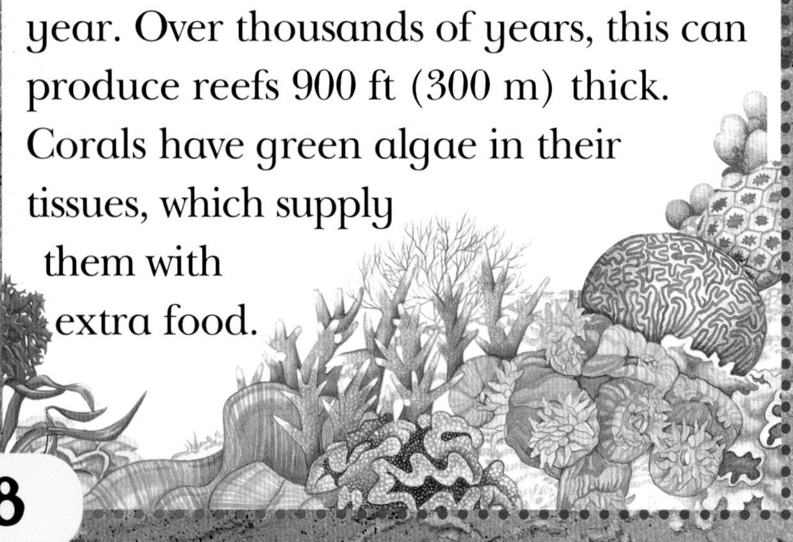

Many reefs are under threat. Soil washed into the sea can smother the polyps. Warmer water temperatures can cause bleaching and may kill polyp colonies. Rising sea levels can "drown" colonies, ending their growth.

those without.

zoom in on...

Mass spawning

All the corals on a large reef spawn on the same night at the same time. Males and females release their eggs and sperm into the water, creating an upside-down "snowstorm" in the sea. Individual sperm and eggs unite, settle on the reef, and grow into a new generation of corals.

An atoll is a ring of coral reef growing on the shoulders of an extinct volcano that has been eroded (worn down) by the sea. Of the 261 known atolls, most are in the Pacific Ocean, where the volcanoes arose originally from weaknesses in the Earth's crust called "hotspots."

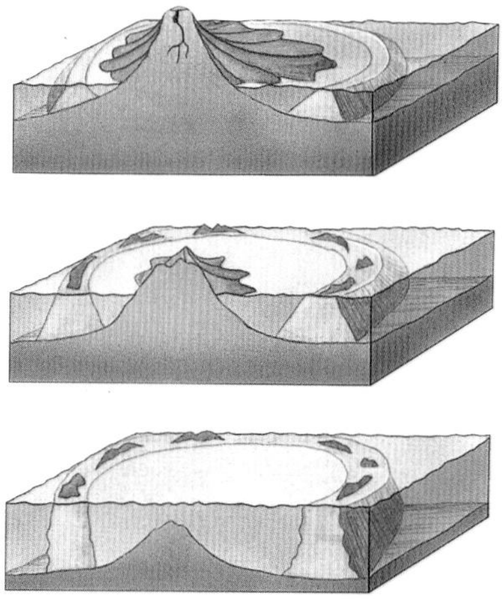

Corals grow around a volcanic island.

The volcano erodes and a lagoon forms.

The volcano sinks and the reef remains with small, sandy islands on top.

Awesome facts

The giant clam can be over 100 years old. It is so strong that you might not be able to break free if its two shells closed on your arm or leg.

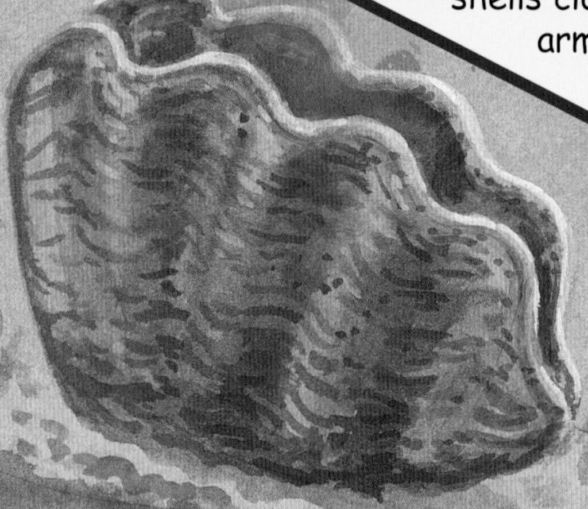

Seahorses have prehensile tails with which they hang onto corals, sponges, and seaweeds. The male gives birth to the youngsters, having taken the eggs from the female into a special pouch.

Q: How does the ferocious moray eel catch its prey?

A: It hides in a hole in the coral by day and emerges at night to hunt. Its jaws are equipped with rows of needle-sharp teeth with which it catches small fish. It competes with whitetip reef sharks, which are also night-time feeders.

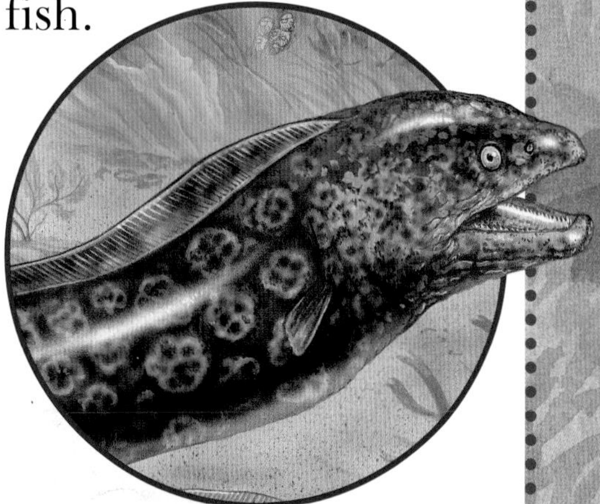

Like a knight in armor, the mantis shrimp hides in the coral, defending its territory. Some species have a pair of hammerlike limbs that can crack the shell of a crab or the exoskeleton of a rival. Others have sharply pointed spears with which they stab their victims.

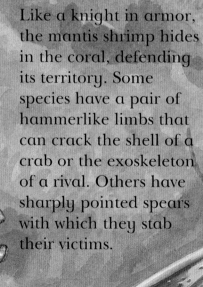

Life on the reef

The reef is like a fortress city, with high-rise corals and sandy avenues. Marine animals come here to feed, escape predators, and have their babies. Residents often have spines or special armor for protection as they hide among the sharp-edged corals. Some small crabs even live inside the corals to escape marauding sea stars.

The crown-of-thorns sea star is the villain of the reef. It snacks on coral polyps, engulfing them with its stomach and digesting them where they stand. One sea star can eat 6 sq yards (5 sq m) of coral a year. It may have more than 20 arms and is covered with spines. Its only foe is the giant triton, a marine snail.

Tropical fish

There are more species of fish on coral reefs than anywhere else in the oceans—up to 200 species in a small area. Fish that live on the shallow sea bed are dull-colored to blend in with the sandy bottom, while those in the coral are camouflaged to match the bright colors there.

Boxfish have an outer boxlike casing of armored plates, like a turtle's shell. This makes them poor swimmers. They feed by standing on their head and squirting water into the sand to reveal small worms and crustaceans.

Clownfish live among the deadly tentacles of the sea anemone, yet are never harmed. Somehow they build up immunity—probably a thick layer of mucus on their skin that shields them from the sting cells. They keep the sea anemone free of food debris and, in return, the sea anemone protects them from larger predatory fish.

zoom in on...

Camouflage

The stonefish lies on the sea bed where it resembles a stone. It waits for small fish to swim close to its mouth, then gulps them in. It is impossible to see, so is also dangerous to people. If you step on it, its dorsal fin spines puncture your foot and inject poison into the wounds.

The triggerfish has a stout spine that can be erected and locked in place. This lets it lodge in nooks and crannies, where it cannot be moved. It is also covered with armored scales. It likes to feed on sea urchins, which it flips over to attack the unprotected underside.

Triggerfish

The whitetip reef shark can bend its dorsal fin in order

The most spectacular gatherings of sharks are the huge schools of up to 500 scalloped hammerheads that swim back and forth off reefs and islands. They are mostly females and do not feed during the day. At dusk, they hunt in twos or threes away from the reef.

How many hammerheads can you see?

By day, whitetip reef sharks rest on the sandy bottom around coral reefs, but at night they go hunting in packs. If a shark detects a movement from a crevice, it dives in headfirst, then the rest of the gang follows. It has a wedge-shaped head and tough skin, which lets it bulldoze its way through coral.

to squirm into crevices.

Tropical sharks

Many shark species live around coral reefs. Each has its own resting and hunting places. Whitetip reef sharks stay on the reef flats, gray reef sharks patrol the drop-off into deep water, and blacktip reef sharks hunt in the shallow, sandy lagoons.

zoom in on...

Cleaner fish

At special places on the reef, whitetip reef sharks have dead skin and parasites removed by cleaner fish. Body stripes show they are cleaners, and they signal that they are open for business with a special dancc. The sharks also adopt a certain posture to say they are ready.

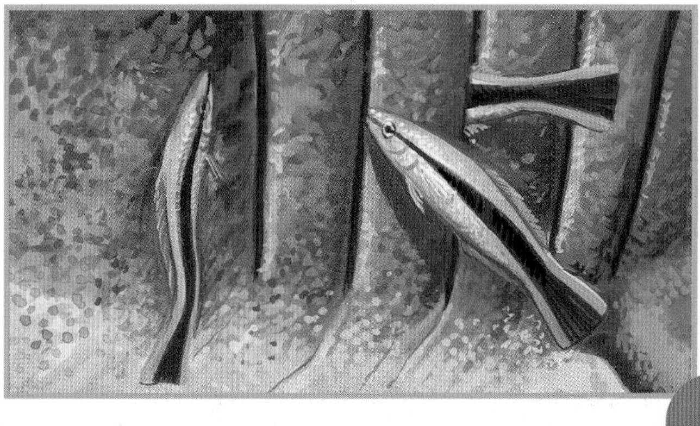

Awesome facts

Gray reef sharks perform a curious stiff-bodied threat display to warn off intruders. They swim in a figure-eight pattern with nose up and pectoral fins down.

Sea snakes

Many tropical sea snakes are dull-colored and relatively harmless, but those with bright stripes and blotches are usually venomous. The open ocean yellow-bellied sea snake has black and yellow stripes that say, "I'm dangerous."

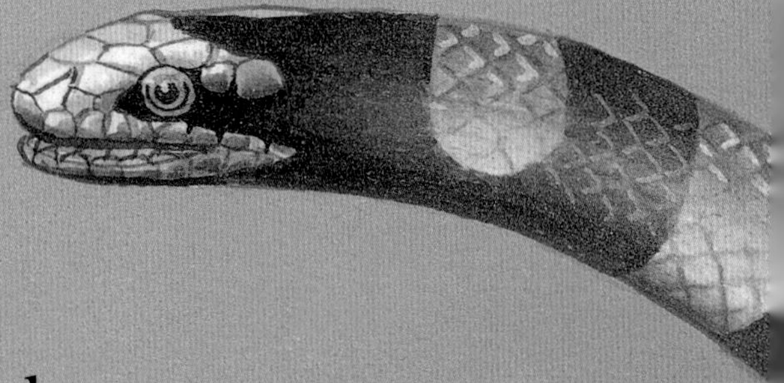

zoom in on...

Sea snake rafts

Sailors have reported seeing huge floating islands of intertwined sea snakes. In 1932, sailors in the Malacca Straits reported a raft of sea snakes stretching more than 60 miles (100 km). Others have been spotted off Pakistan and Vietnam.

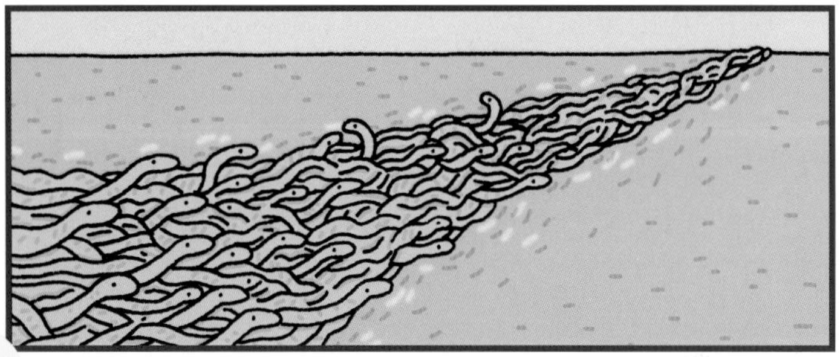

The banded sea snake lives in shallow water off the tropical Asian coasts. It hunts at night and eats mainly eels. Its fangs are short, but the toxin is powerful. Unlike land snakes, which follow their dying victims using smell, sea snakes must kill their prey instantly or they risk losing it in a crevice or having it snatched by another predator.

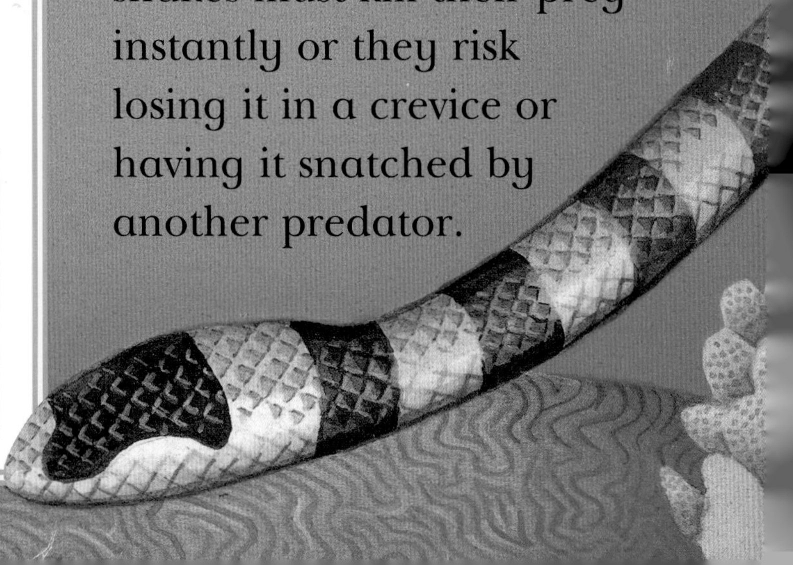

but is harmless to humans.

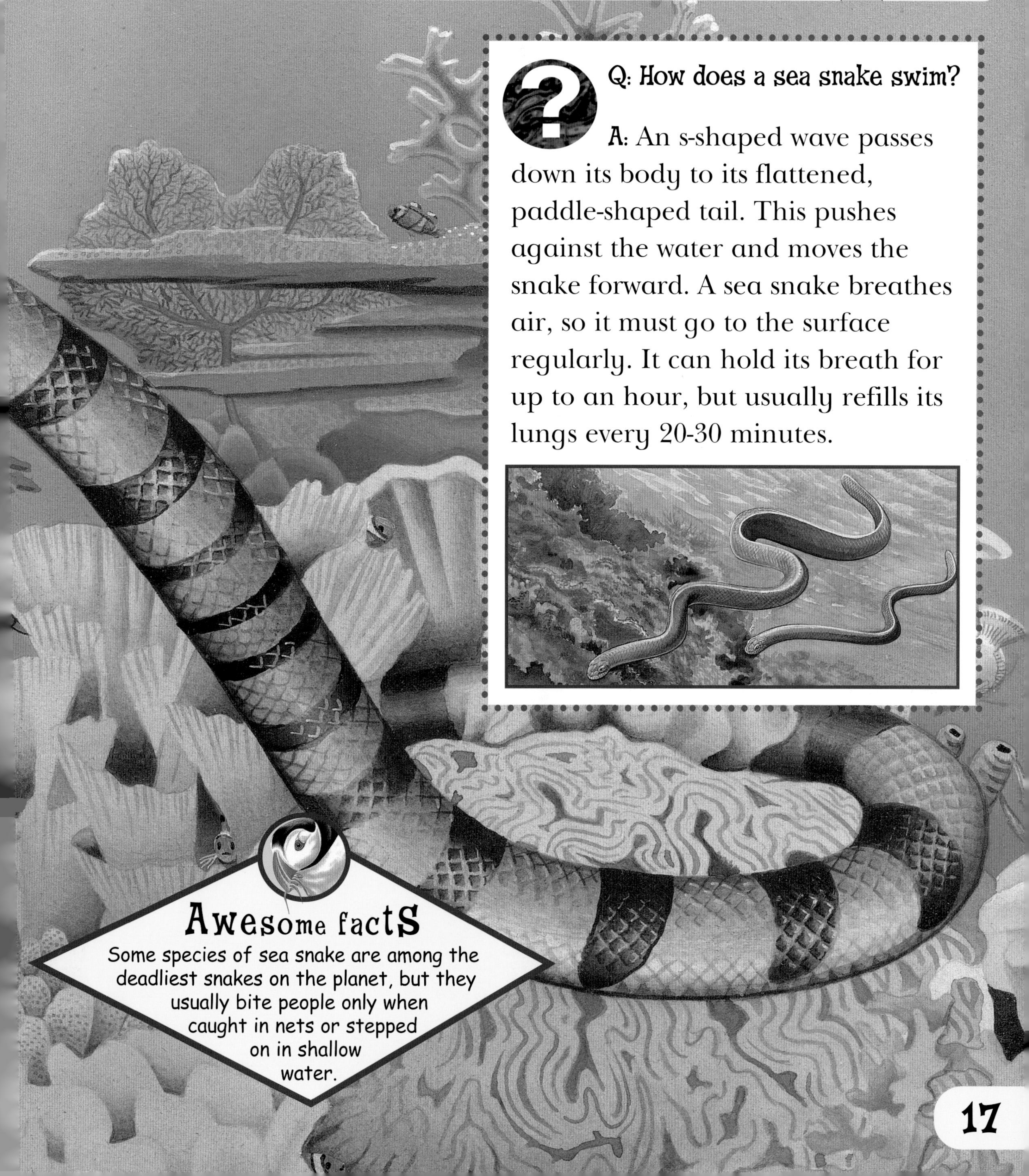

Q: How does a sea snake swim?

A: An s-shaped wave passes down its body to its flattened, paddle-shaped tail. This pushes against the water and moves the snake forward. A sea snake breathes air, so it must go to the surface regularly. It can hold its breath for up to an hour, but usually refills its lungs every 20-30 minutes.

Awesome facts

Some species of sea snake are among the deadliest snakes on the planet, but they usually bite people only when caught in nets or stepped on in shallow water.

Hawksbill turtle shells were once used to make eye-glass

Q: Where do sea turtles breed?

A: Sea turtles mate in the sea and lay their eggs on land. The female crawls up onto a beach—usually the beach where she hatched years before. She digs a hole, lays her eggs, covers over the hole with sand, then returns to the sea. She sighs frequently on land, as her heavy body is not supported by water and it pushes down on her vital organs. A couple of months later, the hatchlings emerge. They instinctively know the way to the sea, but as they travel across the sand, many are eaten by vultures, seabirds, large lizards, and ghost crabs.

Most turtles, like this green turtle, have a hard carapace, or shell, on the back and a plastron, or shield, on the belly. These consist of bones in the skin, which are covered by scutes, or scales. The massive leatherback is the exception. It has bony plates embedded in its leathery skin and has no scutes.

frames, but now these turtles are protected.

Tropical turtles

There are seven species of sea turtle—green, hawksbill, loggerhead, flatback, Kemp's ridley, olive ridley, and leatherback—most of which live or breed in the tropics. The leatherback is the largest, and can be over 6 feet (2 m) long. It lays its eggs in the tropics, but migrates to cooler seas to find the jellyfish on which it feeds. Green turtles also travel long distances.

Awesome facts

Sea turtles can deposit up to 200 spherical eggs at a time, returning two to three times during the nesting period to lay more clutches.

Lifesavers in Australia protect themselves from box jellyfish stings

Tropical jellyfish

Jellyfish live in most seas, including tropical seas. Most species have a jellylike bell with dangling, stinging tentacles. They move by jet propulsion—the bell contracts to push water downward, which pushes the jellyfish up. Although they consist mainly of water, jellyfish can be deadly. Just brushing against the tentacles of a box jellyfish could kill you.

Q: How does a jellyfish sting?

A: Each tentacle is covered with thousands of tiny stinging cells. Inside each cell is a poisonous dart that shoots out when its trigger is touched by prey. The dart's release is the fastest known movement in nature. When the dart strikes, it pumps poisons into the victim.

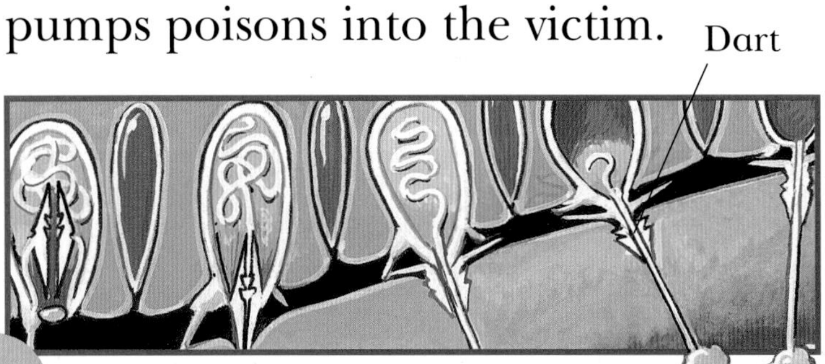

How many spotted jellyfish can you count?

zoom in on...

By-the-wind sailor

The by-the-wind sailor is a disklike relative of jellyfish that floats on the surface of the sea. It has a tiny sail set at an angle and is blown along while searching for food. It falls prey itself to the violet sea snail, which floats at the surface with the aid of a raft of air bubbles.

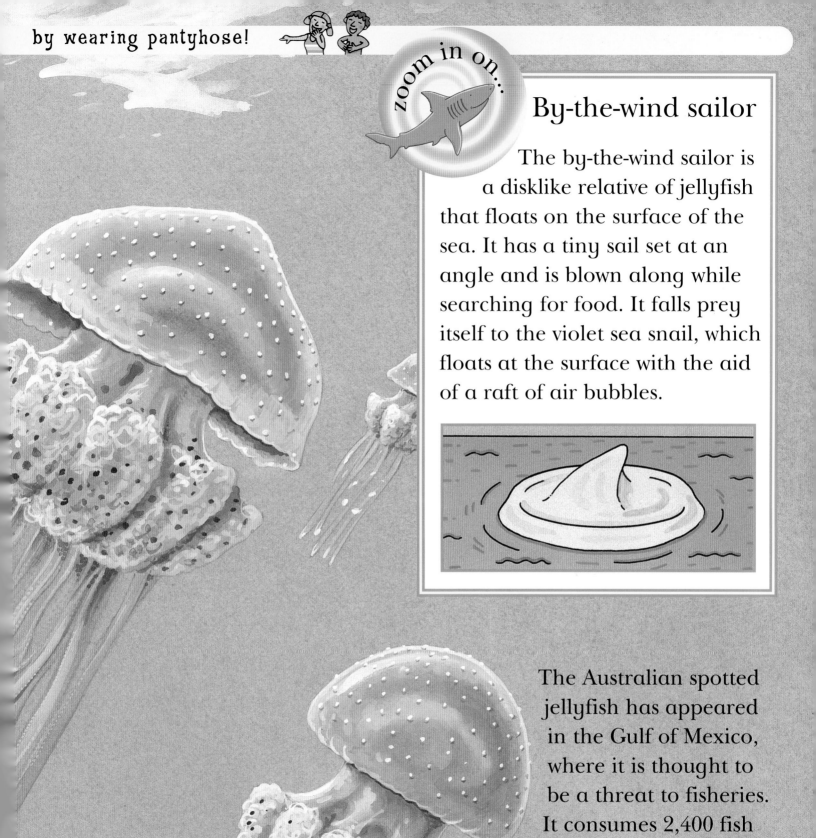

The Australian spotted jellyfish has appeared in the Gulf of Mexico, where it is thought to be a threat to fisheries. It consumes 2,400 fish eggs a day and can grow to be the size of a basketball.

Sea grass has true roots and flowers under the sea.

Sea plants

Tropical seas have green, brown, and red seaweeds like those in cooler waters, but they also have wide meadows of sea grass and amazing plants like mangroves. Both sea grass and mangroves grow in shallow waters protected from ocean swells and storms. The tangle of mangrove roots provides a nursery area for crustaceans, baby fish, and sharks.

Sea grass is food for sea cows and green turtles.

A red tide is the prolific growth of red algae called dinoflagellates. The growth can be so great that the sea is stained red. The algae produce poisons. If shellfish, such as mussels or oysters, eat the algae, the poison is concentrated in their tissues. People who then eat these shellfish are poisoned.

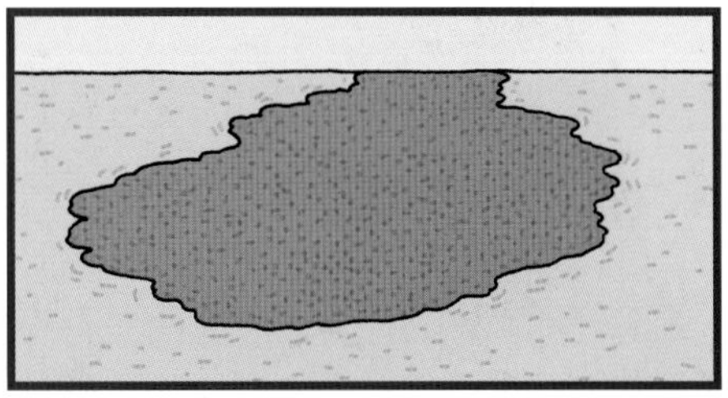

Mangroves dominate tropical salty tidal mudflats. Their stems and roots provide hiding places for baby fish and sharks. They also have their own marine animals, such as fiddler crabs.

The fairy tern doesn't have a nest—it balances and incubates

Tropical seabirds

Tropical seabirds have various hunting techniques, from e white-bellied sea eagle that snatches prey from the urface, to the brown booby that plunges underwater to pursue fish. Diving birds must be careful not to be grabbed by tiger sharks or other predators lurking just below the surface.

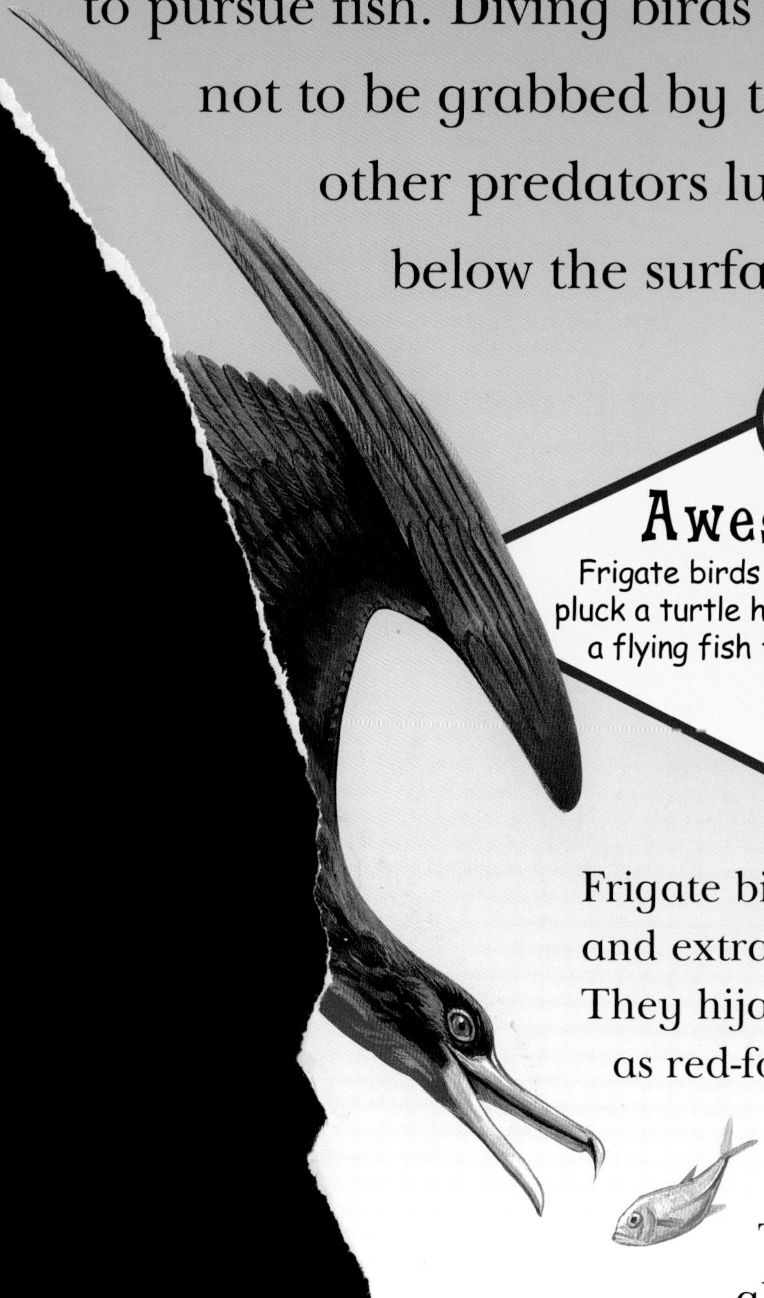

Awesome facts

Frigate birds are so agile that they can pluck a turtle hatchling from the beach or a flying fish from the sea without stopping.

Frigate birds are aerial pirates and extraordinary acrobats. They hijack other birds, such as red-footed boobies, forcing them to regurgitate and drop their food. They then steal it—all in midair.

zoom in on...

Pipefish

Pipefish are long and thin, and closely resemble the seaweeds in which they hide[redacted] They feed on plankton. Male [redacted] female partners greet each [redacted] every morning. One mo[redacted] female transfers her e[redacted] male and he looks [redacted] allows the female [redacted] produce anothe[redacted]

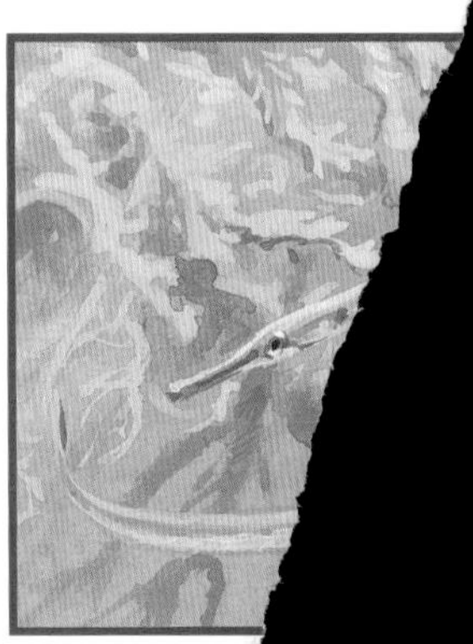

by wearing pantyhose!

By-the-wind sailor

The by-the-wind sailor is a disklike relative of jellyfish that floats on the surface of the sea. It has a tiny sail set at an angle and is blown along while searching for food. It falls prey itself to the violet sea snail, which floats at the surface with the aid of a raft of air bubbles.

The Australian spotted jellyfish has appeared in the Gulf of Mexico, where it is thought to be a threat to fisheries. It consumes 2,400 fish eggs a day and can grow to be the size of a basketball.

Sea grass has true roots and flowers under the sea.

Sea plants

Tropical seas have green, brown, and red seaweeds like those in cooler waters, but they also have wide meadows of sea grass and amazing plants like mangroves. Both sea grass and mangroves grow in shallow waters protected from ocean swells and storms. The tangle of mangrove roots provides a nursery area for crustaceans, baby fish, and sharks.

Sea grass is food for sea cows and green turtles.

A red tide is the prolific growth of red algae called dinoflagellates. The growth can be so great that the sea is stained red. The algae produce poisons. If shellfish, such as mussels or oysters, eat the algae, the poison is concentrated in their tissues. People who then eat these shellfish are poisoned.

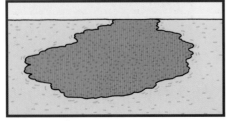

Mangroves dominate tropical salty tidal mudflats. Their stems and roots provide hiding places for baby fish and sharks. They also have their own marine animals, such as fiddler crabs.

its egg on a branch.

The pelican has an enormous bill that it uses like a net to catch fish. Different species of pelican fish in different ways. When the brown pelican (above) spots a school of fish, it folds its wings and plunges below the surface. It later regurgitates the food for its chicks.

White pelicans feed in a group. Several birds take up a u-shaped formation on the sea's surface. They then plunge their heads below to drive fish into the shallows, where they collect them in their enormous pouches.

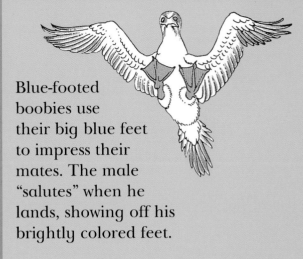

Blue-footed boobies use their big blue feet to impress their mates. The male "salutes" when he lands, showing off his brightly colored feet.

The male then displays by pointing to the sky and opening wide his wings.

The female responds, rattling her wings and raising her head.

The ritual ends in a strutting dance, with the pair bowing to each other.

Early sailors may have thought that sea cows were

Marine mammals

Several species of dolphin, whale, seal, and sea cow live in tropical seas. The slow-swimming Indo-Pacific humpback dolphin hunts coastal waters. Large whales have their calves in tropical waters, but go together to polar seas to feed. Only Bryde's whale stays to feed on fish near tropical reefs.

zoom in on...

Manatees

Manatees are sea cows that can be 16 ft (5 m) long. They browse on sea grasses in the Caribbean and can eat more than 110 lb (45 kg) in one day! The upper lip is divided and closes like pliers on the plants. Manatees are now rare.

Awesome facts

Some species of dolphin have more than 200 teeth, but they use them only to grasp prey, which they then swallow whole, usually headfirst.

The tropical spinner dolphin of the open ocean gathers in schools of 100 or more. It gets its name from its habit of leaping clear of the water and spinning its body three or four times during a single leap. This is thought to be a means of communication.

Q: How do dolphins hunt?

A: A group of dolphins hunts in a line, each scanning the sea ahead with its echolocation system. High-frequency sounds are emitted from its forehead, then its lower jaw picks up the echoes that bounce off prey such as fish or squid. This enables the dolphin to locate its food.

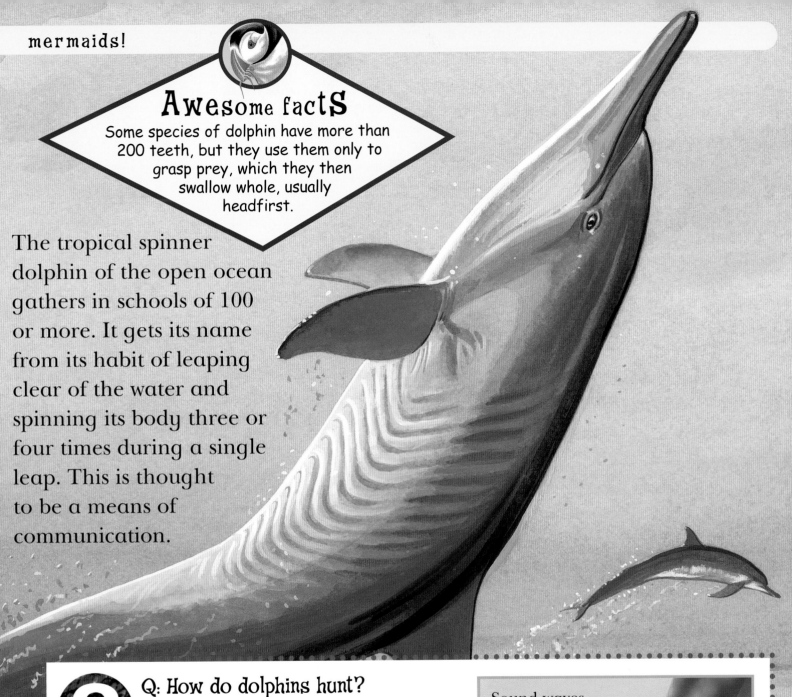

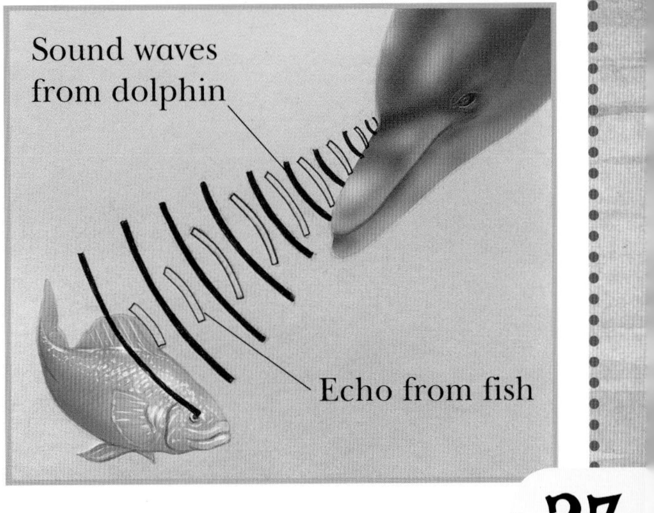

Flying fish travel in groups just

Extra-weird wildlife

The tropical seas play host to both giant and miniature dragons. There are also crabs that come onto land and climb trees, fish that fly, marine iguanas that sneeze salt, fish that change sex, and worms that look like Christmas trees!

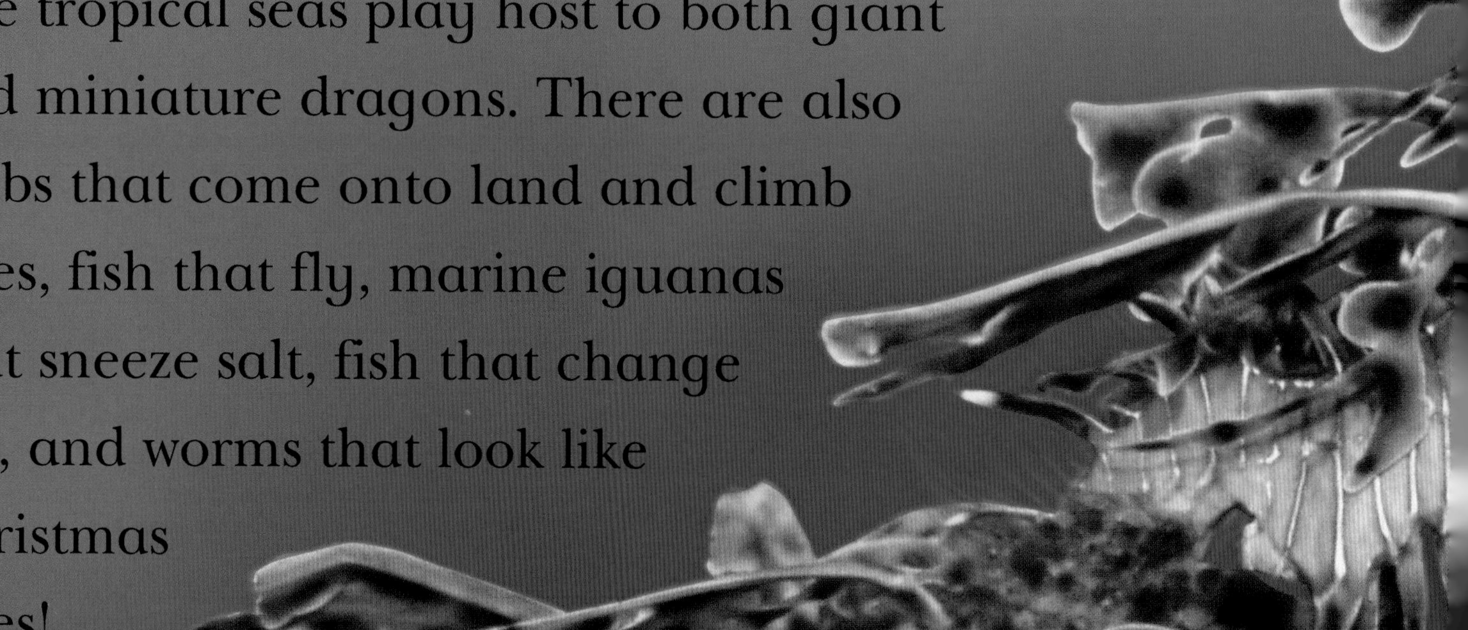

Most crocodile species live in fresh water, but saltwater crocodiles, or "salties," are the world's largest and most powerful living reptiles. They frequent coastal areas and estuaries and are very dangerous. They swim by sweeping their immense tail from side to side, but come onto land to deposit their eggs in piles of vegetation.

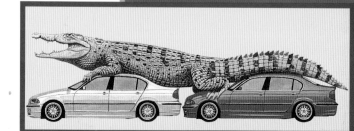

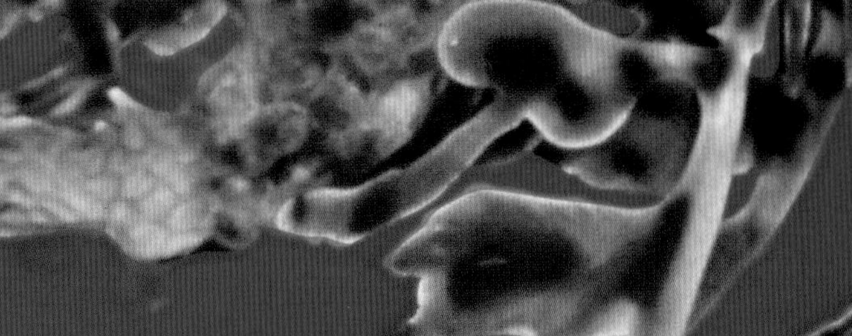

below the sea's surface.

The leafy sea dragon is a type of seahorse with trailing weedlike appendages that look like seaweed. It is camouflaged in this way to blend in with the kelp seaweed in which it lives. It swims very slowly and is found off the Australian coast.

zoom in on...

Flying fish

Flying fish have enlarged pectoral fins that act as wings. The fish swims fast to the surface, beats with its tail, holds out its fins, and glides. A single flight can be more than 100 yards (100 m), helping it to escape from larger attackers.

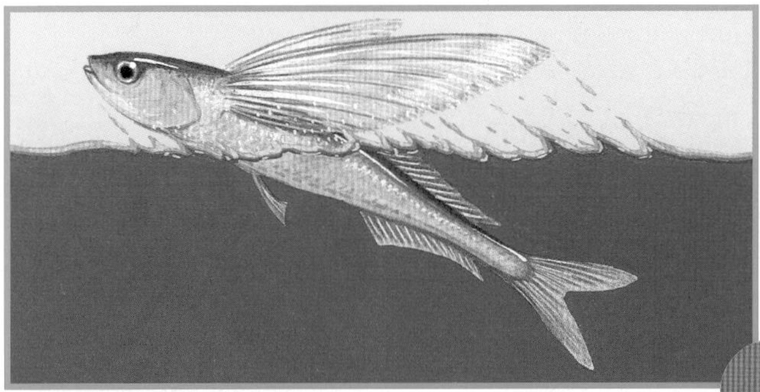

Awesome facts

One saltwater crocodile killed in the Norman River in Australia was an amazing 28.34 ft (8.64 m) long—that's as long as two family cars (see left)!

Endangered in the tropics

Tropical seas, like all seas, have their casualties. In Central America, for example, Kemp's ridley turtle has become the world's most endangered sea turtle. There were 40,000 in 1947, but there are only 1,200 mature females left today. They are confined to a single nesting beach in Costa Rica.

The green turtle has suffered at the hands of egg collectors, soup makers, fisheries, and developers who build hotels on turtle nesting beaches. Now, many are growing potentially dangerous tumors. Nobody knows the cause.

The Caribbean monk seal once lived in the Caribbean islands and was the first large animal from the New World to be reported by the explorer Christopher Columbus. It is now extinct, killed for its meat and oil. Two other tropical or semi-tropical seal species survive, but are endangered.

Glossary

algae
Simple plantlike living things, most of which live in water.

bleach
To lose color and turn white.

camouflage
The way in which an animal hides from predators or stalks prey by blending in with its surroundings.

crustacean
A hard-bodied animal with no backbone but with many hanging body parts that have a variety of functions, from eating to walking.

endangered
Describes an animal which may become extinct if the reasons for its decline continue.

exoskeleton
The jointed, hard outer skeleton of crustaceans, insects, and spiders.

extinct
Describes a species that has not been seen in the wild for 50 years or more.

immunity
The ability of an animal to resist something, such as a disease or poison.

larva
A young stage in certain animals' development that looks nothing like the adult form. Plural: larvae.

migration
The movement of animals to and from particular parts of the ocean, such as feeding and breeding grounds.

mucus
A jellylike secretion produced by the skin or by the lining of tubes in the body.

parasite
An animal (or plant) that lives on or in another animal (or plant), often causing it harm.

plankton
Tiny plants and animals that float at the surface of the sea or a lake.

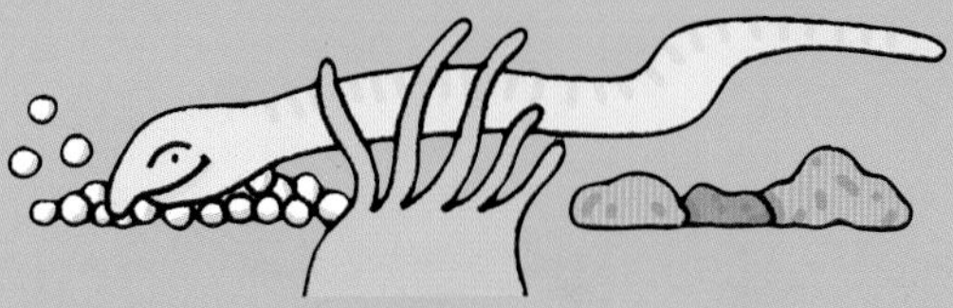

polyp
An animal with no backbone that has a saclike body with a single opening (its mouth). Coral is formed by individual sea anemonelike polyps.

predator
An animal that hunts and eats other animals.

prehensile
Describes a body part, like a tail, that is able to grasp.

prey
An animal that is hunted and eaten by other animals.

regurgitation
The act of bringing up food from the stomach.

species
Animals that resemble one another closely and are able to breed together.

toxin
A poisonous substance.

tumor
An abnormal growth of new tissue.

Index